PHOTOSYNTHESIS

For Curious Little Minds

Plants. Have you ever seen plants eat?

No.

Plants are smart. They can make their own sugar!

How?

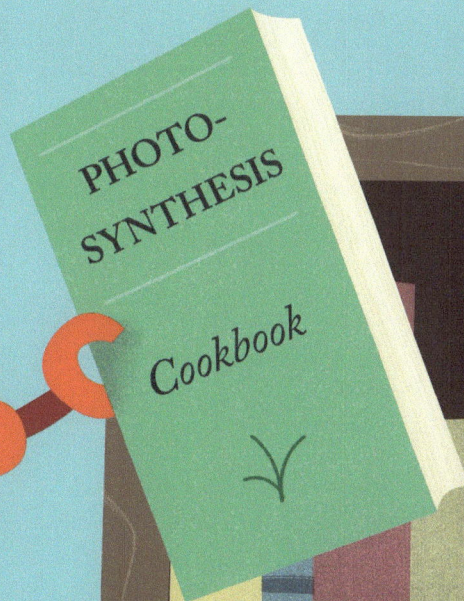

First, Chef Chloroplast collects sunlight. Sunlight is made of a rainbow of color balls called photons.

PHOTONS

Chef Chloroplast stores the red and blue photons in a special bag called chlorophyll. He throws away the green photons. This is why leaves look green.

Second, Chef Chloroplast collects carbon dioxide found in the air. It goes into the leaf through tiny holes called stomata.

CARBON DIOXIDE

STOMATA

Third, Chef Chloroplast collects water found in the ground. Water enters the plant from the roots. Straws, called xylem, let the water travel to the leaf.

WATER

XYLEM

What are those bubbles?

The bubbles are oxygen which the plants let off into the air. You breathe in oxygen to help your body work properly.

OXYGEN

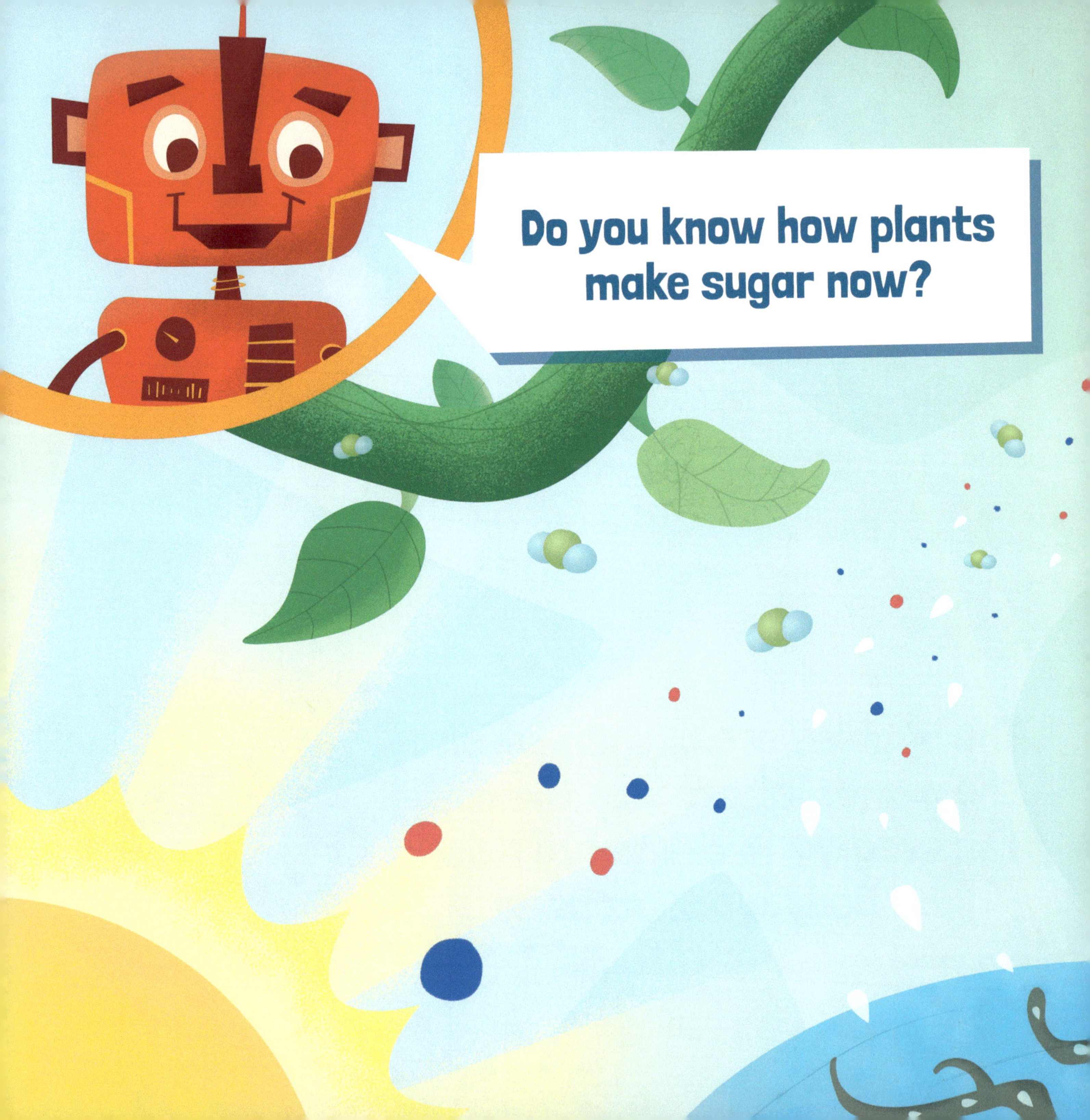
Do you know how plants make sugar now?

Yes! By using photosynthesis. Chef Chloroplast, found inside plants, mixes sunlight, carbon dioxide, and water to make sugar and oxygen.

Can you help Chef Chloroplast find and color the right ingredients needed for photosynthesis?

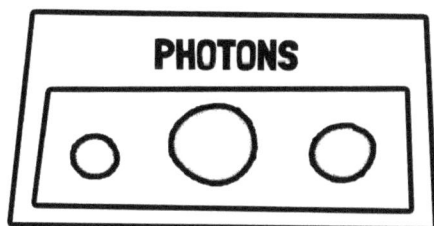

PHOTONS

WATER

CARBON DIOXIDE

OXYGEN

SUGAR

Do you remember which color photon balls Chef Chloroplast throws away to give plants their color?